Let's Learn About Matter

FLEXIBILITY

Rebecca Kraft Rector

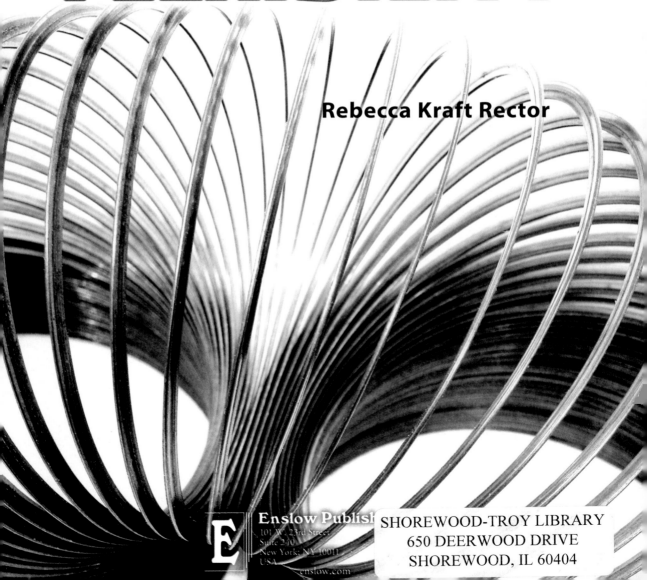

Enslow Publishing
101 W. 23rd Street
Suite 240
New York, NY 10011
USA
enslow.com

SHOREWOOD-TROY LIBRARY
650 DEERWOOD DRIVE
SHOREWOOD, IL 60404

Words to Know

atom A tiny bit of matter.

brittle Easily cracked or broken.

chemical Having to do with chemistry.

chemistry The science that deals with properties of matter and how it forms and changes.

gas A kind of matter that has no permanent shape, like air.

liquid A kind of matter that can move freely, like water.

physical Having to do with being able to be touched or seen.

properties The qualities or features of something.

solid A kind of matter that is firm and keeps its shape.

Contents

WORDS TO KNOW 2
WHAT IS MATTER? 5
COMMON FORMS OF MATTER 7
PROPERTIES OF MATTER 9
THE PROPERTY OF FLEXIBILITY 11
FLEXIBILITY AND FORCE 13
FLEXIBILITY AND SOLID MATTER 15
WHAT'S FLEXIBLE? 17
HOT AND COLD CHANGES 19
WET AND DRY CHANGES 21
ACTIVITY: CAN IT BEND? 22
LEARN MORE 24
INDEX 24

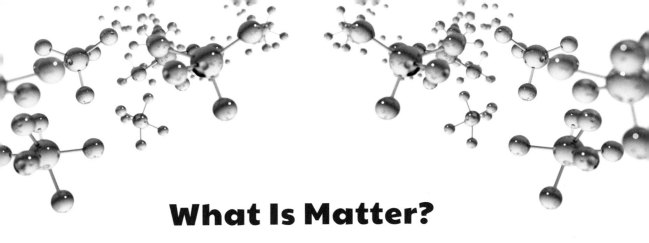

What Is Matter?

Matter is everything around you. All things are made of matter. Tiny bits of matter are called atoms. Atoms join together to make molecules.

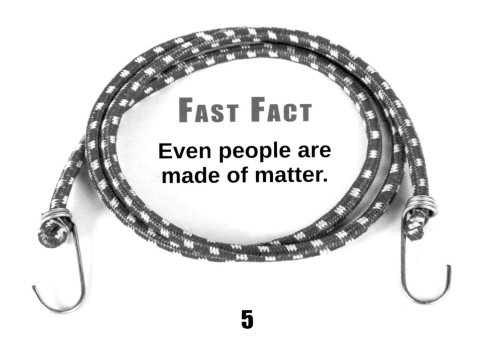

Fast Fact

Even people are made of matter.

All three forms of matter are present here in the air, the water, and the towel.

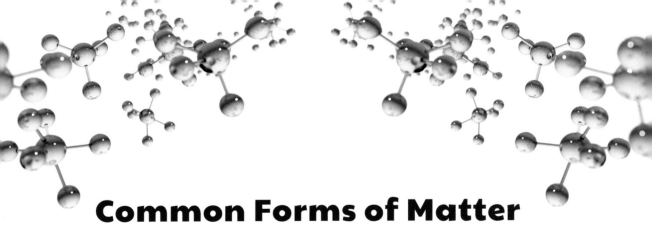

Common Forms of Matter

Matter has different forms. Matter can be solid. A tree is a solid. Matter can be liquid. A waterfall is a liquid. Matter can be a gas. Helium is a gas.

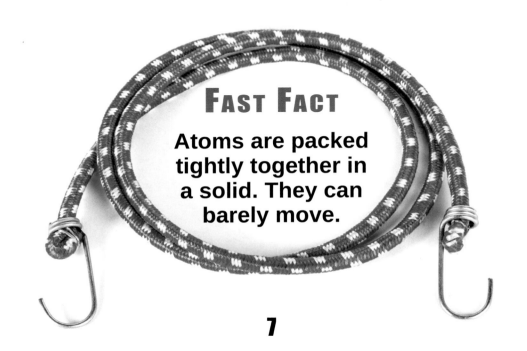

Fast Fact

Atoms are packed tightly together in a solid. They can barely move.

Living things have many physical properties. They come in different sizes, colors, and shapes.

Properties of Matter

Properties tell about matter. Physical properties tell how it acts, looks, and feels. Is it big or small? Hard or soft? Chemical properties let matter change. An example is how easily something catches on fire.

A basketball net is flexible. It moves and changes shape as the basketball goes through it.

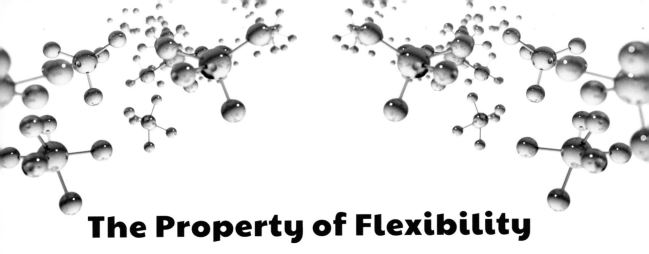

The Property of Flexibility

Flexibility is a physical property. It is how much something can bend without breaking. It helps tell how something acts. You can see flexibility. You can feel it, too.

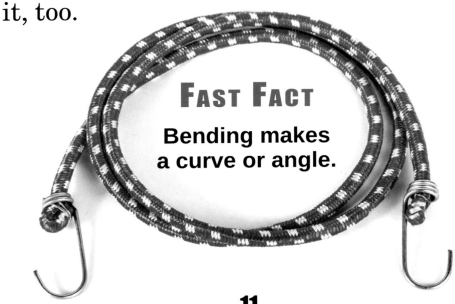

Fast Fact

Bending makes a curve or angle.

We can see paper's flexibility when we fold it.

Flexibility and Force

Flexible matter can change its shape. It can be bent or folded. Force must be used. Pushing is a force. You must push your hands together to bend something.

Fast Fact

Thin pieces are usually easier to bend than thick pieces.

A garden hose does not break when it is wound up. It is flexible.

Flexibility and Solid Matter

Solid matter may be flexible. It may bend without breaking. Some solid matter is not flexible. It may be stiff or brittle. Brittle matter often breaks.

Fast Fact

You can sort materials by flexibility.

What's Flexible?

Will it bend? Or will it break? These questions help tell what is flexible. Paper, clay, and cloth are flexible. Most plastics and rubber bend easily. Pencils break.

Fast Fact

Twisting shows flexibility.

Horseshoes are made by heating up steel and then bending it into shape.

Hot and Cold Changes

Some materials can bend if they are heated. Steel is one of them. Cold can make materials brittle. Frozen plastic may break.

Dry noodles break when you bend them. When they are placed in a pot of water, they become flexible.

Wet and Dry Changes

Some materials bend more easily when they are wet. Wet straw is more flexible than dry straw. Wet noodles are flexible. Dry noodles are brittle.

Activity
Can It Bend?

MATERIALS
Water
Popsicle sticks
Container
Cup or glass

Let's find out if wood is flexible.

Step 1: Fill the container with water.

Step 2: Soak popsicle sticks in the water overnight. Or you could soak the sticks in boiling water for 30 minutes. Have an adult help.

Step 3: Carefully bend the wet sticks around the outside of the cup.

Step 4: Place the bent sticks inside the cup to dry overnight.

Step 5: Decorate the sticks and wear as wrist bands.

Wood is flexible when wet! Wood keeps its shape when dry.

See what happens to wooden popsicle sticks when you soak them in water.

Learn More

Books

Diehn, Andi. *Matter: Physical Science for Kids*. White River Junction, VT: Nomad Press, 2018.

Enz, Tammy. *Bend It!* Chicago, IL: Heinemann Raintree, 2018.

Owen, Ruth. *Let's Investigate Everyday Materials*. New York, NY: Bearport, 2017.

Websites

BBC Bitesize
www.bbc.com/bitesize/articles/zx8hhv4
Read about how to identify materials.

BBC Science
www.bbc.co.uk/schools/scienceclips/ages/7_8/characteristics_materials.shtml
Discover properties of matter with this fun game.

Index

atoms, 5, 7
bending, 11, 13, 15, 17, 19, 21
breaking, 11, `15, 17, 19
brittle, 15, 19, 21
chemical properties, 9
cold, 19
flexibility, 11, 13, 15, 17, 21
force, 13
gas, 7
heat, 19
liquid, 7
matter, 5, 7, 8, 15
physical properties, 9, 11
solid, 7, 15
wet changes, 21, 22-23

Published in 2020 by Enslow Publishing, LLC.
101 W. 23rd Street, Suite 240, New York, NY 10011

Copyright © 2020 by Enslow Publishing, LLC.

All rights reserved.

No part of this book may be reproduced by any means without the written permission of the publisher.

Library of Congress Cataloging-in-Publication Data

Names: Rector, Rebecca Kraft, author.
Title: Flexibility / Rebecca Kraft Rector.
Description: New York : Enslow Publishing, 2020. | Series: Let's learn about matter | Audience: K to grade 4. | Includes bibliographical references and index.
Identifiers: LCCN 2018047005| ISBN 9781978507531 (library bound) | ISBN 9781978509078 (pbk.) | ISBN 9781978509085 (6 pack) Subjects: LCSH: Matter—Properties—Juvenile literature. | Flexure—Juvenile literature. Classification: LCC QC173.36 .R4275 2020 | DDC 530.4—dc23 LC record available at https://lccn.loc.gov/2018047005

Printed in the United States of America

To Our Readers: We have done our best to make sure all website addresses in this book were active and appropriate when we went to press. However, the author and the publisher have no control over and assume no liability for the material available on those websites or on any websites they may link to. Any comments or suggestions can be sent by e-mail to customerservice@enslow.com.

Photos Credits: Cover, p. 1 charles taylor/Shutterstock.com; p. 4 Jacek Chabraszewski/Shutterstock.com; p. 6 TravnikovStudio/Shutterstock.com; p. 8 Eric Isselee/Shutterstock.com; p. 10 Melinda Nagy/Shutterstock.com; p. 12 AnastasiaNi/Shutterstock.com; p. 14 Tharin Sinlapachai/Shutterstock.com; p. 16 AlexMaster/Shutterstock.com; p. 18 Noska Photo/Alamy Stock Photo; p. 20 maradon 333/Shutterstock.com; p. 23 feelphoto2521/Shutterstock.com; interior design elements (bungee cord) Natalia Rezanova/Shutterstock.com, (molecules) 123dartist/Shutterstock.com.